Sodium carbonate –
a versatile material

Compiled by Ted Lister

ROYAL SOCIETY OF CHEMISTRY

Sodium carbonate – a versatile material

Compiled by Ted Lister

Edited by Colin Osborne and Maria Pack

Designed by Imogen Bertin and Sara Roberts

Published by The Royal Society of Chemistry

Printed by The Royal Society of Chemistry

Copyright © The Royal Society of Chemistry 2000

For further information on other educational activities undertaken by The Royal Society of Chemistry write to:

The Education Department
The Royal Society of Chemistry
Burlington house
Piccadilly
London W1V OBN

Information on other Royal Society of Chemistry activities can be found on its websites:
www.rsc.org
www.chemsoc.org

ISBN 0–85404–924–X

British Library Cataloguing in Publication Data.

A catalogue for this book is available from the British Library.

Cover pictures reproduced by kind permission of Brunner Mond.

RS•C

Contents

RS•C

Introduction

This booklet presents learning material based on the manufacture and uses of sodium carbonate made by the Solvay (ammonia-soda) process. It is the result of a Learning Material Workshop organised by The Royal Society of Chemistry in conjunction with The Institute of Materials and The Worshipful Company of Armourers and Brasiers. The workshop was held at the Brunner Mond Company, Northwich, Cheshire.

A group of chemistry teachers spent a day at Brunner Mond and was given a presentation by the company on various aspects of the Solvay (ammonia-soda) process for the manufacture of sodium carbonate and sodium hydrogencarbonate. This was followed by a tour of the plant. The following day was spent brainstorming and drafting the material which is presented here in edited form.

The teachers involved were:

Paul Butlin, Ipswich High School;

Janice Chubb, The Dame Alice Harpur School, Bedford;

David Earnshaw, Ysgol Y Berwyn, Bala;

Keith Huggins, Aylesbury High School; and

Barry McFarland.

Acknowledgements

The Royal Society of Chemistry thanks Brunner Mond and in particular Martin Ashcroft, David Bain, Alan Duncan, Pip Wright and Bob Yould. It also thanks Chris Baker for setting up the workshop. It also thanks The Institute of Materials and The Worshipful Company of Armourers and Brasiers for their support of the workshop.

The material

The booklet includes teacher's notes and material to photocopy as follows:

Part 1. Manufacturing sodium carbonate – an overview for teachers
An overview for teachers of sodium carbonate and sodium hydrogencarbonate manufacture (especially by the Solvay process) and of the uses of these products. It includes some unusual details of the process and anecdotes which could be used by teachers to enhance their own teaching.

Part 2. Making sodium carbonate
A worksheet for 11–13 year old students. It includes some simple practical work as well as an account of the Solvay process and questions on the process and raw materials.

Part 3. Manufacturing sodium carbonate
A worksheet for 14–16 year old students with an account of the Solvay process and questions on the process and raw materials.

Part 4. Manufacturing sodium carbonate by the Solvay process
An account of the Solvay process based on 1 above but with some of the detail, that which is intended specifically for teachers, removed. It is aimed at post-16 students but would also be accessible to more able pre-16 students.

RS•C

Part 5. The thermodynamics and equilibria involved in the Solvay process for the production of sodium carbonate
A worksheet for post-16 students with questions on the thermodynamics and acid-base aspects of the Solvay process and the uses of the products. It could be used independently but would ideally be used after students have read part 4.

Using the material

None of the material assumes any prior knowledge of the Solvay process. Parts 3 and 5 are entirely self-contained and require no teacher input. They are therefore suitable as homework exercises or as work to be tackled in the event of teacher absence.

The Solvay process is not a core part of current syllabuses and specifications but the aim of the worksheets is to get students to apply chemical principles in an unfamiliar context, not for them to learn details of this process.

RS•C

Teacher's notes, answers and practical requirements

1. Manufacturing sodium carbonate – an overview for teachers

None required

2. Making sodium carbonate

Experimental work – The lime cycle

Apparatus (per group)
▼ Bunsen burner
▼ Clean tin lid
▼ Heatproof mat
▼ Dropping pipette
▼ 100 cm^3 beaker
▼ Test tube and rack
▼ Small filter funnel
▼ Filter paper
▼ Glass rod
▼ Tongs or tweezers
▼ Drinking straw.

Chemicals (per group)
▼ One small lump of calcium carbonate (a marble chip), ideally labelled 'calcium carbonate' or 'limestone' rather than 'marble chips'
▼ One piece of Universal Indicator paper

It is the responsibility of teachers to carry out an appropriate risk assessment for all experimental work.

Uses of products

Sodium carbonate
▼ Glass – *eg* bottles, TV screens, windows, windscreens
▼ Detergents – dishwashers, washing powders
▼ Metal refining
▼ Cement
▼ Tanning
▼ Producing polymers
▼ Water treatment
▼ Extracting sugar from sugar beet
▼ Foods – flavouring

RS•C

Sodium hydrogencarbonate
▼ Baking powder
▼ Water treatment
▼ Cleaning buildings (as an alternative to sandblasting)
▼ Producing foams
▼ Animal food additives – *eg* to strengthen the shells of eggs
▼ Buffers
▼ Fire fighting
▼ Detergents
▼ Antacids, toothpastes, mouthwashes

Calcium chloride
▼ Gelling agents in food
▼ Brewing
▼ Cheese making
▼ Oil and gas drilling
▼ Treating effluent
▼ Fruit spraying

Answers to questions

1. Calcium carbonate → calcium oxide + carbon dioxide

2. Calcium oxide + water → calcium hydroxide

3. Calcium hydroxide + carbon dioxide → calcium carbonate + water
   The cloudiness is calcium carbonate.

4. $CaCO_3 \rightarrow CaO + CO_2$
   $CaO + H_2O \rightarrow Ca(OH)_2$
   $Ca(OH)_2 + CO_2 \rightarrow CaCO_3 + H_2O$

5. (a) Raw materials: salt, coke, calcium carbonate.
   (b) Final products: sodium carbonate, sodium hydrogencarbonate, calcium chloride.
   (c) Intermediates: carbon dioxide, ammonium chloride, ammonia, sodium hydrogencarbonate.

6. Exothermic: $C + O_2 \rightarrow CO_2$
   $CaO + H_2O \rightarrow Ca(OH)_2$
   $CO_2 + H_2O + NH_3 \rightarrow NH_4HCO_3$
   (although students are unlikely to get this one)

   Endothermic: $CaCO_3 \rightarrow CO_2 + CaO$
   $2NaHCO_3 \rightarrow Na_2CO_3 + H_2O + CO_2$
   (although students are unlikely to get this one)

7. Any suitable uses from the lists above.

8. Sodium hydrogencarbonate.

9. Ammonia.

10. If some of the ammonium chloride were sold, this would result in less calcium chloride being produced. However, as some ammonia would now be used up, this would have to be added.

RS•C

3. Manufacturing sodium carbonate

Answers to questions

1. Various. The list might include windows (houses, cars), bottles, jars, TV and computer screens, light bulbs, drinking glasses, mirrors.

2. Salt is present in the area. (The many names ending in -wich in the area – Northwich, Nantwich, Middlewich, for example – testify to the importance of salt; the ending -wich is derived from Anglo-Saxon.) Limestone is available close by (as is coal, from which the coke is made).

3. (a) Na^+ and CO_3^{2-}.
   (b) Na^+ from sodium chloride and CO_3^{2-} from calcium carbonate.

4. The salt extraction process has little environmental impact. All that is visible above ground are a number of small pumping stations each about the size of a garden shed. On the other hand, quarries are large, difficult to hide and may be considered an eyesore for many years after they have finished working (unless landscaped).

5. (a) and (e).

6. Ammonia is an intermediate (a catalyst might also be an acceptable term at this level) as it is neither a raw material nor a product.

7. Calcium chloride.

8. (a) The reaction regenerates the ammonia for re-use in reaction (c).
   (b) The price for which calcium hydroxide and ammonium chloride could be sold relative to the price for which ammonia is bought. Ammonia becomes in effect a reactant, rather than an intermediate.

9. Burning coke (an exothermic reaction) provides the heat energy to decompose the limestone (an endothermic reaction).

10. The carbon dioxide is used in the process.

4. Manufacturing sodium carbonate by the Solvay process

None required

5. The thermodynamics and equilibria involved in the Solvay process for the production of sodium carbonate

Answers to questions

1. (a) Yes it is.
   (b) $+ 20$ kJ mol^{-1}
   The value is the same as that quoted for the overall reaction.

2. Coke is an impure form of carbon produced by heating coal in the absence of air for about 18 hours. Moisture, a variety of volatile organic compounds (coal tar) and ammonia are driven out of the coal during this process.

3. -393 kJ mol^{-1}

4. +179 kJ mol^{-1}

5. 2

6. Burning coke produces carbon dioxide as the only main product. This is used in the Solvay process.

7. Far less earth has to be moved; it is safer; it has less environmental impact.

8. As we move down from the top of the tower, more reaction has taken place, more heat is given out and the temperature rises. Further down the tower, the cooling water removes much of this heat and the temperature drops again.

9. The rate of formation of the product and the temperature.

RS•C

10. $2NaHCO_3(s) \rightarrow Na_2CO_3(s) + H_2O(l) + CO_2(g)$ $\qquad \Delta H = +92$ kJ mol^{-1}

11. A: $CaCO_3 \rightarrow CO_2 + CaO$ and $C + O_2 \rightarrow CO_2$
    B: $CaO + H_2O \rightarrow Ca(OH)_2$
    C: $2NaHCO_3 \rightarrow Na_2CO_3 + H_2O + CO_2$
    or the equivalent word equations
    a: calcium oxide
    b: carbon dioxide
    c: sodium hydrogencarbonate
    d: calcium hydroxide

12. (a) Acid: H_2O; base: NH_3
    (b) Acid: NH_4^+; base: OH^-

13. (a) The solution will be alkaline. The carbonate ion will accept a proton from a water molecule.
    $$CO_3^{2-}(aq) + H_2O(l) \rightleftharpoons HCO_3^-(aq) + OH^-(aq)$$
    leaving free OH^- ions in solution.

    (b) For the equilibrium:
    $$CO_3^{2-}(aq) + H_2O(l) \rightleftharpoons HCO_3^-(aq) + OH^-(aq)$$

    $$2.08 \times 10^{-4} = \frac{[HCO_3^-(aq)]_{eq}\,[OH^-(aq)]_{eq}}{[CO_3^{2-}(aq)]_{eq}}$$

    At equilibrium, $[HCO_3^-(aq)]_{eq} = [OH^-(aq)]_{eq}$
    and, since we are dealing with a weak base, in a 1 mol dm^{-3} solution of sodium carbonate $[CO_3^{2-}(aq)]_{eq} \approx 1$ mol dm^{-3}

    $$2.08 \times 10^{-4} = \frac{[OH^-(aq)]_{eq}^2}{1}$$

    $[OH^-(aq)]_{eq} = 0.01443$ mol dm^{-3}
    More OH^- ions are formed from the equilibrium:
    $$HCO_3^-(aq) + H_2O(l) \rightleftharpoons H_2CO_3(aq) + OH^-(aq)$$
    for which

    $$2.22 \times 10^{-8} = \frac{[H_2CO_3(aq)]_{eq}\,[OH^-(aq)]_{eq}}{[HCO_3^-(aq)]_{eq}}$$

    At equilibrium, $[H_2CO_3(aq)]_{eq} = [OH^-(aq)]_{eq}$
    We have calculated above that $[HCO_3^-(aq)]_{eq} = 0.01443$ mol dm^{-3} and, since we are dealing with a weak base, in a 0.01443 mol dm^{-3} solution of hydrogencarbonate ions

    $[HCO_3^-(aq)]_{eq} \approx 0.01443$ mol dm^{-3}

    $$2.22 \times 10^{-8} = \frac{[OH^-(aq)]_{eq}^2}{0.01443}$$

    $[OH^-] = 1.79 \times 10^{-5}$ mol dm^{-3}
    So the total $[OH^-]$ from both equilibria is
    $1.79 \times 10^{-5} + 0.01443 = 0.0145$ mol dm^{-3}
    Since

    $[H^+(aq)]_{eq} \times [OH^-(aq)]_{eq} = 1.0 \times 10^{-14}$ mol^2 dm^{-6}

    $[H^+(aq)]_{eq} \times 0.0145 = 1.0 \times 10^{-14}$

RS•C

$[H^+(aq)]_{eq} = 6.9 \times 10^{-13}$ mol dm^{-3}

pH = 12.2

14. The essential point is to select an indicator that will change sharply at the second end point. Thymol blue's red to yellow change is probably the most suitable with methyl orange as as a more familiar alternative.

    Also look for appropriate detail of making up the solution and the titration procedure itself.

    If it is desired to allow students to actually carry out a titration, a sample of sodium carbonate contaminated with a known percentage of sodium chloride could be made up.

    The analysis of this mixture could give practice in making up solutions in volumetric flasks and other aspects of volumetric analysis.

RS•C

Part 1
Manufacturing sodium carbonate

An overview for teachers

RS·C

Manufacturing sodium carbonate – an overview for teachers

Sodium carbonate (Na_2CO_3) is used by many different industries as a raw material, and about one million tonnes is produced each year in the UK – all of it by the Brunner Mond Company in Northwich, Cheshire. Also produced in smaller quantities is sodium hydrogencarbonate ($NaHCO_3$) as well as calcium chloride ($CaCl_2$) – a by-product, a little of which can be sold.

Industrially, sodium carbonate is usually referred to as 'soda ash' and is produced and sold in two grades:

▼   'light ash' – a fine powder; and

▼   'heavy ash' – which has a bigger particle size and is more dense, making it less bulky to transport.

Sodium hydrogencarbonate is used in:

▼   water treatment;

▼   as an additive in food and drinks – *eg* baking powder;

▼   for blowing foams such as expanded polystyrene;

▼   in pharmaceutical products as an antacid;

▼   in personal care products such as toothpaste; and

▼   as an additive in animal feeds.

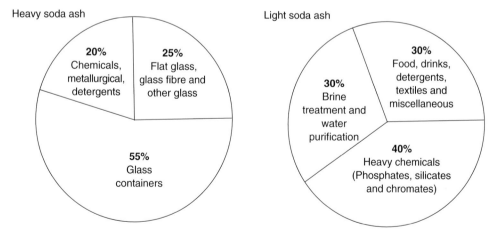

Fig 1 The uses of sodium carbonate
Exact percentages will vary with economic and social factors

Figure 1 gives an approximate breakdown of the uses of light ash and heavy ash but these are subject to change depending on a number of social and economic factors. For example, in a recession, fewer cars and houses are built, which reduces the demand for glass. Importing of alcoholic drinks from the continent due to more liberal customs regulations has led to a decrease in the number of glass bottles made in the UK and thus a drop in demand for heavy ash.

RS•C

The Solvay process

This process has been used for making sodium carbonate and sodium hydrogencarbonate since the late 19th century when it began to replace the Leblanc process. No more effective process has been found. The Solvay process uses salt (sodium chloride) to provide the sodium ions and limestone (calcium carbonate) for the carbonate ions in the sodium carbonate.

Salt and limestone are cheap and plentiful raw materials. Salt is found in underground deposits in Cheshire. It is extracted by solution mining as brine which is then pumped to the site and treated to precipitate out calcium and magnesium ions. The calcium carbonate is quarried as limestone near Buxton in Derbyshire and arrives on site by rail.

The reaction below seems to be the obvious way to prepare sodium carbonate from sodium chloride and calcium carbonate.

$$2NaCl(aq) + CaCO_3(s) \rightarrow Na_2CO_3(aq) + CaCl_2(aq) \qquad \Delta H = +20 \text{ kJ mol}^{-1}$$
$$\Delta G = +60 \text{ kJ mol}^{-1}$$

Unfortunately salt and limestone do not react together. The value of ΔG shows that the equilibrium lies well to the left, so an indirect route must be used. In fact calculating K_c from the expression $\Delta G = -RT \ln K_c$ gives a value of

$K_c \approx 1 \times 10^{-10} \text{ dm}^3 \text{ mol}^{-1}$. Furthermore, the overall reaction is endothermic and so a significant input of heat energy is required. This is provided by burning coke which thus becomes the third raw material. This is transported to the site by road.

The key reaction is that between sodium chloride solution and carbon dioxide in the presence of ammonia. This is a reversible reaction forming ammonium chloride and sodium hydrogencarbonate. It occurs in 25 metre-high Solvay towers where a downward flow of ammonia dissolved in brine meets an upward flow of carbon dioxide.

$$NaCl(aq) + NH_3(aq) + H_2O(l) + CO_2(g) \rightleftharpoons NH_4Cl(aq) + NaHCO_3(s)$$

Energy for the overall process is provided by burning coke.

$$C(s) + O_2(g) \rightarrow CO_2(g) \qquad \Delta H = -393 \text{ kJ mol}^{-1}$$

The heat generated by this reaction is used to decompose the calcium carbonate to provide the carbon dioxide.

$$CaCO_3(s) \rightarrow CaO(s) + CO_2(g) \qquad \Delta H = +178 \text{ kJ mol}^{-1}$$

The ΔH values show that two moles of calcium carbonate can be decomposed with the heat energy from one mole of coke, and this approximate ratio is used in practice. Both reactions produce carbon dioxide for the process.

Returning to the key reaction.

$$NaCl(aq) + NH_3(aq) + H_2O(l) + CO_2(g) \rightleftharpoons NH_4Cl(aq) + NaHCO_3(s)$$

Sodium hydrogencarbonate is much less soluble than ammonium chloride at low temperature and crystallises out. The equilibrium in the Solvay tower reaction thus moves to the right. The reaction is exothermic ($\Delta H = -79 \text{ kJ mol}^{-1}$), so the tower is cooled to keep the temperature at its base down to about 25 °C. The sodium hydrogencarbonate is filtered out and heated to form sodium carbonate ('light ash').

$$2NaHCO_3(s) \rightarrow Na_2CO_3(s) + CO_2(g) + H_2O(g)$$

The carbon dioxide is recycled.

The calcium oxide from the decomposition of the limestone is slaked with water to form calcium hydroxide.

RS•C

$$CaO(s) + H_2O(l) \rightarrow Ca(OH)_2(s)$$

This is used to regenerate the ammonia in the distiller.

$$Ca(OH)_2(s) + 2NH_4Cl(aq) \rightarrow CaCl_2(aq) + 2NH_3(aq) + 2H_2O(l)$$

The ammonia is recycled.

The overall process is shown in a simplified form in *Figure 2* and more pictorially in *Figure 3*. It operates as two cycles, an ammonia cycle and a carbon dioxide cycle. In theory, no ammonia is used up; it is all recycled. In practice, a little is required to make up losses.

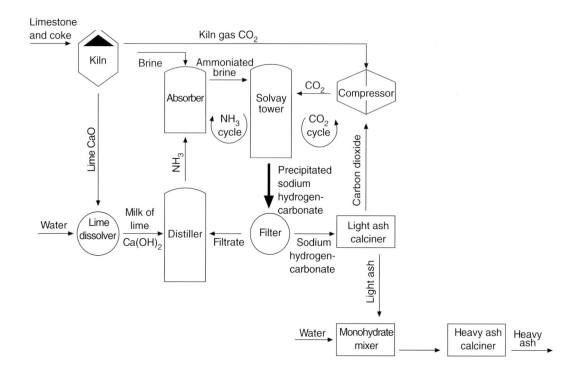

**Fig 2 A simplified flow diagram for the Solvay process
showing the ammonia and carbon dioxide cycles**

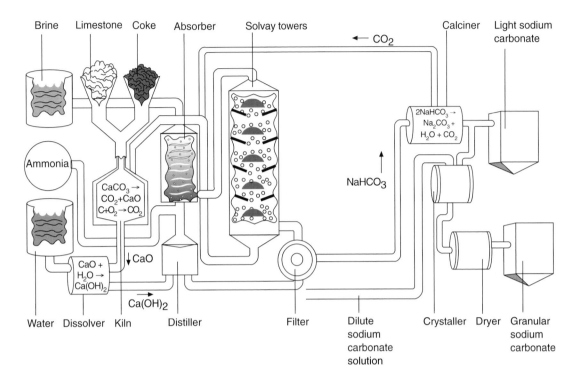

Fig 3 The Solvay process

The Solvay process manufactures three different products – light sodium carbonate (light ash), granular sodium carbonate (heavy ash) and refined sodium hydrogencarbonate.

Light sodium carbonate (light ash)

This is made by taking filtered sodium hydrogencarbonate and heating it. This drives off water and carbon dioxide, which can be recycled. The product is a very fine white powder.

Granular sodium carbonate (heavy ash)

Light sodium carbonate is made into a slurry with water, to form sodium carbonate monohydrate. This is then heated to produce the anhydrous form as much larger crystals. These crystals have a particle size similar to that of sand so that the two make a homogeneous mixture which is important for glass making – the major use for heavy ash.

Refined sodium hydrogencarbonate

Crude sodium hydrogencarbonate is filtered and decarbonated and dehydrated by heating to give sodium carbonate. This is dissolved in water and the resulting solution is filtered to remove impurities. Highly pure sodium hydrogencarbonate crystals are formed by reacting the filtered solution with carbon dioxide. These crystals are then centrifuged and dried in a carbon dioxide atmosphere. The product is one of the purest industrial chemicals, and can be added to foods and pharmaceutical products.

The main by-product of the Solvay process is calcium chloride. A little of this can be sold for use in refrigeration, curing concrete and as a suspension in oil drilling. The bulk of calcium chloride is disposed of in the nearby river Weaver.

RS•C

Alternative sources of sodium carbonate

So-called natural ash can be obtained from lakes containing alkaline brine and sodium sesquicarbonate dihydrate, (trona) ($Na_2CO_3.NaHCO_3.2H_2O$). Such deposits are found in a number of areas of the world including Kenya, Egypt and the US. In Wyoming, US, the deposits are about 500 m below the surface and are accessed by deep mining techniques.

Trona can be converted to soda ash by heating to 500 °C followed by recrystallisation from aqueous solution.

Interesting facts concerning the industrial process

Corrosion of the plant
The ammonia used in the process comes, not from the Haber process as might be expected, but as a by-product of making coke for this and other processes. (The Solvay process, of course, pre-dates the Haber process.) This ammonia contains traces of both hydrogen sulfide and cyanide ions. Most of the Solvay process plant is made of cast iron and those parts of it which are kept at a high temperature (such as the distillers) tend to corrode in contact with a hot, aqueous solution containing chloride ions.

However, hydrogen sulfide reacts with the iron to form hard, insoluble iron(II) sulfide which coats the inside of the plant and prevents corrosion. In some locations, where hydrogen sulfide is driven out by boiling, the cyanide ions react to form hexacyanoferrate(II) ions which react further with the iron to give a very hard, dark blue coating of $(NH_4)_2Fe[Fe(CN)_6]$.

When this impure ammonia was replaced by purer Haber process ammonia the distillers corroded severely within a matter of months. Nowadays, the plant uses coal tar ammonia or, if Haber process ammonia is used, small quantities of cyanides and sulfides are added as corrosion inhibitors.

The limestone kilns
The coke and limestone are mixed in a kiln and ignited. While the kiln is in continuous operation, the incoming coke ignites automatically at the temperature of the kiln. The coke used is produced to very detailed specifications. It must be strong enough to withstand being crushed by the limestone when it is tipped into the kiln. It must also have a specified calorific value – ie the amount of heat given out on burning a specified amount of coke – to produce the correct amount of energy. Because it is stored outside in heaps and may absorb rainwater, it is measured by volume and not mass.

Not all of the limestone decomposes in the kiln. After slaking, the unreacted calcium carbonate (called 'backstone') is removed from the dissolver and returned to the kiln. About 6 per cent of the limestone is returned as backstone.

Lighting the kilns
The kilns normally run continuously. However when they do need to be relit (a process which occurs only once every 12–15 years), the process is somewhat arcane. Railway sleepers and rags soaked in creosote are placed in the kiln along with a small amount of coke. A gunpowder charge is used to ignite this and then more coke added as it catches light, producing puffs of smoke from the top of the kiln. This process is irreverently called 'electing the Pope'. Gradually the coke being added is enriched in limestone until the correct proportions are reached.

Hot water effluent
Hot water from the process is put into the nearby river Weaver. This has encouraged the local fish in this area to migrate to the outflow pipe. 30 lb carp have been caught close to the plant.

RS•C

Aggregate tax
The government has proposed (1999) a tax of £5 per tonne on limestone quarried for aggregate – the stone used for road building. If this tax is also levied on limestone used as a chemical (as in the Solvay process), it will affect the cost of sodium carbonate and hence the cost of glass *etc.*

Commodity chemicals and speciality chemicals
Commodity chemicals are those with reasonably high specifications which are sold in thousands of tonnes but at relatively low prices. Speciality chemicals have very high specifications and are sold in much smaller quantities but at higher prices. There is no precise definition of either term. Both sodium carbonate and sodium hydrogencarbonate can be considered as a commodity chemical or a speciality chemical depending on the exact specification in terms of purity, particle size, *etc.*

Further information

Further details of the Solvay process can be obtained from the following sources.

The essential chemical industry (4th ed). York: The Chemical Industry Education Centre,1999.

Sodium carbonate, R D A Woode, *Steam*, issue 7, April 1987.

There is a section on solution mining for salt in Cheshire in the clip *Chemicals from salt* in the video *Industrial chemistry for schools and colleges* , London: RSC, 1999.

This page has been intentionally left blank.

RS•C

Part 2
Making sodium carbonate

Material for 11–13 year old students

Making sodium carbonate

Sodium carbonate is a very useful material. About one million tonnes of it is made each year in the UK – all by the Brunner Mond company in Cheshire. Although it is a white powder, most of it is used in making glass. This involves mixing it with sand and other substances and heating it. To make sodium carbonate, Brunner Mond uses two raw materials:

▼ salt (chemical name sodium chloride) which is found in underground deposits close to their factory; and

▼ limestone (chemical name calcium carbonate), which is quarried in Derbyshire.

The lime cycle

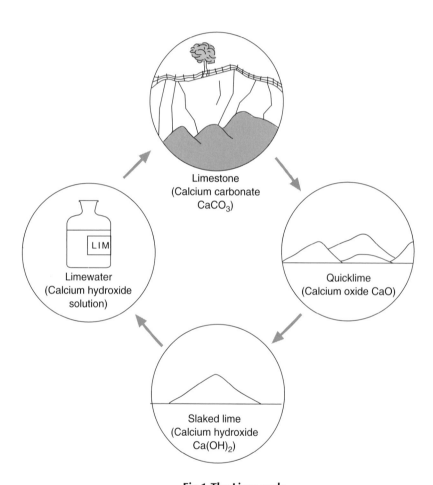

Fig 1 The Lime cycle

Figure 1 shows a cycle of chemical reactions starting with limestone. These turn limestone into other substances and finally back to calcium carbonate. Some of these reactions are used by Brunner Mond when it makes sodium carbonate.

Experiment

You can try some of these reactions yourself. You must wear eye protection.

1. Take a small chip of calcium carbonate and place it on a clean tin lid standing on a heatproof mat. Heat the chip strongly for a few minutes using the hottest part of a roaring Bunsen flame, *Figure 2*. Note the burner air hole should be open, and remember that the hottest part of a Bunsen flame is just beyond the blue cone.

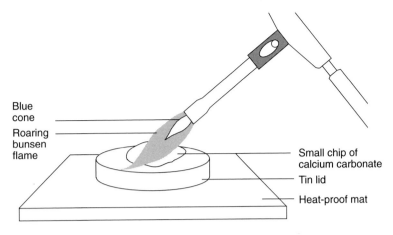

Fig 2

Does the limestone still look the same? Write down anything you observe.

2. Let the limestone chip cool (this may take a few minutes – be patient). Use tongs to place the chip in a small beaker and add water from a dropper one drop at a time *Figure 3.*

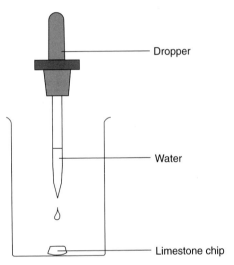

- Dropper
- Water
- Limestone chip

Fig 3

Write down your observations.

3. Now add more water – about 15 cm^3 – *Figure 4,* and stir with a glass rod to make a solution. There will still be a lot of unchanged limestone left.

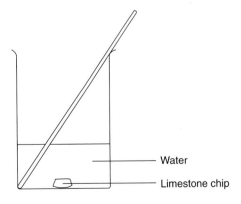

- Water
- Limestone chip

Fig 4

4. Filter your solution into a test tube, *Figure 5*. The solution should be clear. If it is not, filter it again. Use Universal Indicator paper to find the pH of the solution.

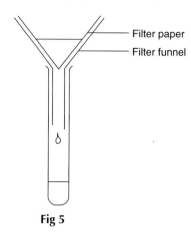

Fig 5

5. Gently blow into your clear solution with a straw, *Figure 6*. Do not suck! Describe what you see.

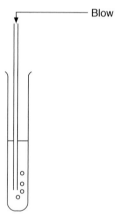

Fig 6

The Solvay process for making sodium carbonate

Brunner Mond uses the carbon dioxide that it gets from heating limestone and reacts it with a solution containing salt (sodium chloride) and ammonia. This process is named after the Belgian chemist, Ernst Solvay who developed it in the 1860s.

At their factory in Cheshire, Brunner Mond does this reaction in huge towers, about the size of six coaches stacked on top of each other. They filter off the cloudy solid, which is called sodium hydrogencarbonate (sometimes called sodium bicarbonate or bicarbonate of soda). They can sell some of this, and the rest is heated to turn it into sodium carbonate.

Sodium hydrogencarbonate and sodium carbonate are used in a number of everyday products and are sometimes labelled as E500.

Try to find out how these compounds are used. Look for any of their names or the code E500 on packs of products at home. You could try food products (in the kitchen), cosmetics and personal care products (in the bathroom cabinet), cleaning products (under the sink) and medicines. Do take care during your search and put things back! During your search you could also look for products containing calcium chloride, which is also made in the same process, and calcium carbonate itself.

You could make a poster with pictures of the products or cut out and stick on the actual labels (only when the packs are empty!).

RS•C

Questions

1. Using *Figure 1* to help you, write a word equation to describe the reaction which occurs when you heat calcium carbonate.

2. Using *Figure 1* to help you, write a word equation to describe the reaction between calcium oxide and water.

3. When you blow into the limewater, you are adddding carbon dioxide from your breath. Write a word equation to describe the reaction which occurs. What is the cloudiness you see?

4. Write symbol equations for the reactions in questions 1–3.

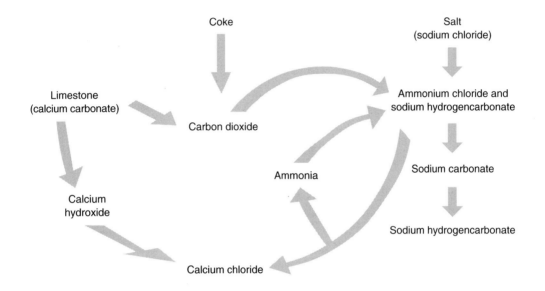

Fig 7

The flow chart in *Figure 7* shows an outline of the Solvay process. Look at the chart and try to answer the following questions.

Questions

5. Mark on a copy of the chart.
    (*a*)   The raw materials.
    (*b*)   The final products.
    (*c*)   Any intermediates (these are chemicals which take part in the process but which are neither raw materials nor products).

    You could use different coloured pens or highlighters for each category.

6. Some of the reactions in the cycle are exothermic (they give out heat) and some are endothermic (heat must be put in to make them go). Use a red highlighter to mark any processes which you know are exothermic and a blue highlighter to mark any which you know are endothermic. Don't worry if you cannot do all the sections. Thinking back to the experiments you did in *The lime cycle* might help.

7. Suggest two uses for each of the products.

8. Which chemical appears both as a product and an intermediate?

9. Try to identify a re-usable intermediate.

10. Ammonium chloride was previously used in some types of batteries. If this type of battery became popular again, the demand for ammonium chloride would increase. What effect might this have on the overall production scheme?

RS•C

RS•C

Part 3
Manufacturing sodium carbonate

Material for 14–16 year old students

Manufacturing sodium carbonate

You might be surprised to find that a million tonnes of sodium carbonate is made and sold in the UK each year. This is about 20 kg (20 bags of sugar) for every man, woman and child in the country. However, it is unlikely that you or your family have bought any at all, although a few people might find a small packet of 'washing soda' under the sink. This is sodium carbonate and is still used by a few people to soften water or to help clean the oven or the drains. You will have used sodium carbonate indirectly, though, as 90 per cent of it is used to make glass.

Question

> 1. Make a list of at least 10 things you have used or bought in the last week which use glass. You can probably think of lots more than 10, so try to make your list as varied as possible.

All the sodium carbonate made in the UK comes from one company – the Brunner Mond company based in Northwich in Cheshire. The company makes sodium carbonate from limestone (calcium carbonate, $CaCO_3$) and salt (sodium chloride, NaCl) using coke (made from coal) as a fuel. The map in *Figure 1* shows where these raw materials are found.

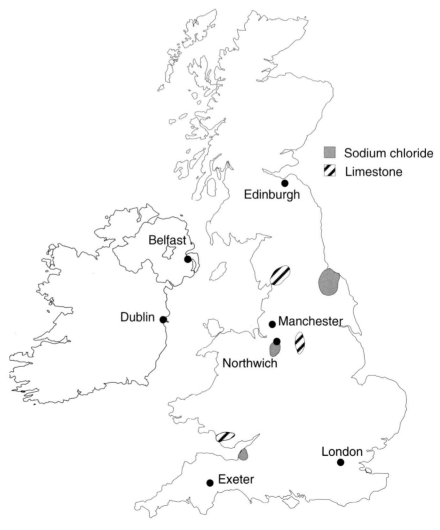

Fig 1 The locations of some mineral deposits in the UK

Figure 2 shows some of the uses of sodium carbonate as well as those of the two raw materials.

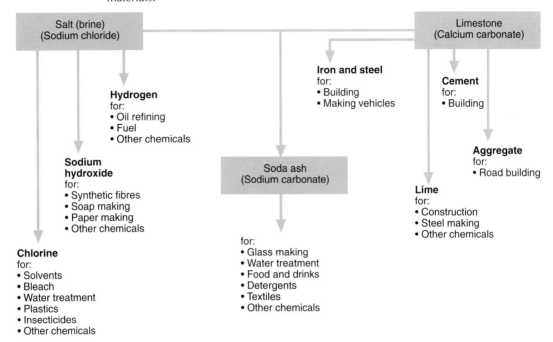

Fig 2 Uses of salt, limestone and sodium carbonate

The manufacture of sodium carbonate is an important use of both salt and limestone.

Salt deposits are found underground, and salt is obtained as brine by 'solution mining' in which water is pumped down a bore hole into the salt beds. It dissolves some of the salt to form brine which is brought to the surface.

Limestone is extracted from quarries.

Questions

2. Look at *Figure 1*. Why do you think that the manufacture of sodium carbonate is based in Northwich in Cheshire?

3. (a) Which ions are present in the compound sodium carbonate, Na_2CO_3?

   (b) The two raw materials used for making sodium carbonate are sodium chloride and calcium carbonate. Which ion is supplied by which raw material?

4. Compare the environmental impact of the different methods used for extracting salt and limestone.

Unfortunately sodium chloride and calcium carbonate do not react directly with each other. However, a sequence of reactions can be used to achieve the same overall result. This sequence can be represented by the equations below.

(a) calcium carbonate $\xrightarrow{\text{Heat}}$ calcium oxide + carbon dioxide

(b) calcium oxide + water → calcium hydroxide

(c) sodium chloride + ammonia solution + carbon dioxide →
       ammonium chloride + sodium hydrogencarbonate

(d) calcium hydroxide + ammonium chloride →
       calcium chloride + ammonia + water

(e) sodium hydrogencarbonate $\xrightarrow{\text{Heat}}$ sodium carbonate + carbon dioxide

The raw materials and products are underlined.

Questions

> 5. Which two reactions are examples of decomposition?
>
> 6. Comment on the role of ammonia in the overall process.
>
> 7. Name a by-product of the process. A by-product is one which is produced by the process but is not really needed.

Even though the ammonia does not get used up in the overall process, it is necessary to top up the supply from time to time.

The process also uses coke (essentially carbon) which is burned in a kiln along with limestone to decompose the limestone into carbon dioxide and calcium oxide. The carbon dioxide is used in reaction (*c*) and the calcium oxide is 'slaked' with water to produce the calcium hydroxide used in reaction (*d*).

Questions

> 8. Reaction (*d*) is not a vital part of the production of sodium carbonate as you can see by looking at the reactants and products.
>
>   (*a*)   Why do you think that the reaction is carried out?
>
>   (*b*)   Could the calcium hydroxide and ammonium chloride be sold as by-products instead of the calcium chloride? What factors determine whether this is an economically sensible alternative?
>
> 9. What is the main purpose of the coke in the furnace with the limestone? (Use the words exothermic and endothermic in your answer.)
>
> 10. What additional role is played by the combustion of the coke?

Part 4
Manufacturing sodium carbonate by the Solvay process

RS•C

An overview of the process for post-16 students

Manufacturing sodium carbonate by the Solvay process

Sodium carbonate is used by many different industries as a raw material and about one million tonnes is produced each year in the UK – all of it by the Brunner Mond Company in Northwich, Cheshire. Also produced in smaller quantities is sodium hydrogencarbonate as well as calcium chloride as a by product, a little of which can be sold.

Industrially, sodium carbonate is usually referred to as 'soda ash' and is produced and sold in two grades:

▼ 'light ash' – a fine powder; and

▼ 'heavy ash' which has a bigger particle size and is more dense, making it more efficient to transport.

Sodium hydrogencarbonate is used in:

▼ water treatment;

▼ as an additive in food and drinks – *eg* baking powder;

▼ for blowing foams such as expanded polystyrene;

▼ in pharmaceutical products as an antacid;

▼ in personal care products such as toothpaste; and

▼ as an additive in animal feeds.

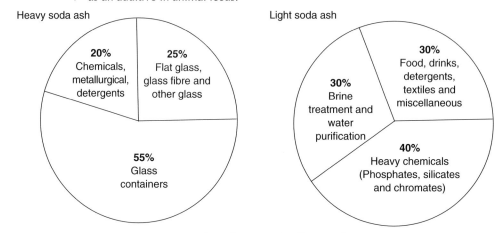

Heavy soda ash

| 20% Chemicals, metallurgical, detergents | 25% Flat glass, glass fibre and other glass |
| 55% Glass containers | |

Light soda ash

| 30% Brine treatment and water purification | 30% Food, drinks, detergents, textiles and miscellaneous |
| 40% Heavy chemicals (Phosphates, silicates and chromates) | |

Fig 1 The uses of sodium carbonate
Exact percentages will vary with economic and social factors

Figure 1 gives an approximate breakdown of the uses of light and heavy ash but these are subject to change depending on a number of social and economic factors. For example in a recession, fewer cars and houses are built, which reduces the demand for glass. Importation of alcoholic drinks from the continent due to more liberal customs regulations has led to a decrease in the number of glass bottles made in the UK and thus a drop in demand for heavy ash.

The Solvay process

This process uses sodium chloride to provide the sodium ions and calcium carbonate for the carbonate ions in sodium carbonate. Salt and limestone are cheap and plentiful raw materials. Salt is found in underground deposits in Cheshire. It is extracted by solution mining as brine which is pumped to the Northwich site and treated to precipitate out calcium and magnesium ions. The calcium carbonate is quarried as limestone near Buxton in Derbyshire and arrives on site by rail.

At first sight, the reaction below seems to be the obvious way to prepare sodium carbonate from sodium chloride and calcium carbonate.

$$2NaCl(aq) + CaCO_3(s) \rightarrow Na_2CO_3(aq) + CaCl_2(aq)$$

Unfortunately salt and limestone do not react together. In fact the reverse reaction actually takes place between calcium chloride and sodium carbonate to give the original starting materials, so an indirect route must be found. Furthermore, the overall reaction is endothermic and so a significant energy input is required. This is provided by burning coke which thus becomes the third raw material. This is transported to the site by road.

The key reaction is that between sodium chloride solution and carbon dioxide in the presence of ammonia. This is a reversible reaction forming ammonium chloride and sodium hydrogencarbonate. It occurs in 25 metre high Solvay towers.

$$NaCl(aq) + NH_3(aq) + H_2O(l) + CO_2(g) \rightleftharpoons NH_4Cl(aq) + NaHCO_3(s)$$

Energy is provided by burning coke, and the heat generated is used to decompose the calcium carbonate to provide carbon dioxide for the process.

$$CaCO_3(s) \rightarrow CaO(s) + CO_2(g)$$

Returning to the key reaction.

$$NaCl(aq) + NH_3(aq) + H_2O(l) + CO_2(g) \rightleftharpoons NH_4Cl(aq) + NaHCO_3(s)$$

At low temperatures the sodium hydrogencarbonate is much less soluble than ammonium chloride and crystallises out. This moves the equilibrium in the Solvay tower reaction to the right. The sodium hydrogencarbonate is filtered out and heated to form sodium carbonate.

$$2NaHCO_3(s) \rightarrow Na_2CO_3(s) + CO_2(g) + H_2O(g)$$

The calcium oxide from the decomposition of the limestone is slaked with water to form calcium hydroxide.

$$CaO(s) + H_2O(l) \rightarrow Ca(OH)_2(s)$$

The calcium hydroxide is used to regenerate the ammonia.

$$Ca(OH)_2(s) + 2NH_4Cl(aq) \rightarrow CaCl_2(aq) + 2NH_3(aq) + 2H_2O(l)$$

The overall process is shown in a simplified form in *Figure 2* and more pictorially in *Figure 3*. It operates as two cycles, an ammonia cycle and a carbon dioxide cycle. In theory, no ammonia is used up; it is all recycled. In practice, a little is required to make up losses.

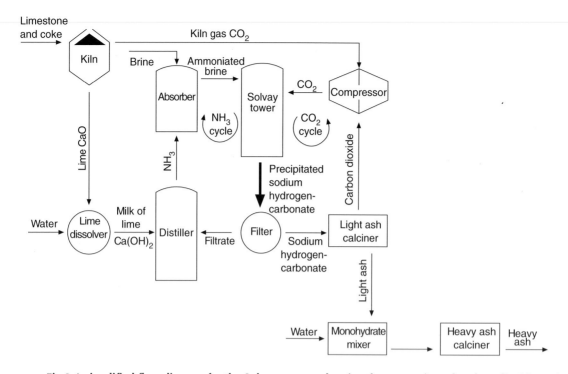

Fig 2 A simplified flow diagram for the Solvay process showing the ammonia and carbon dioxide cycles

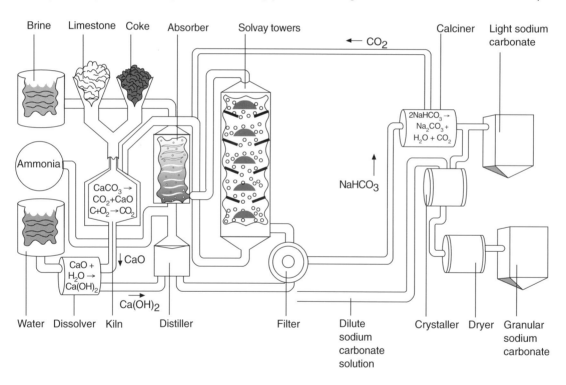

Fig 3 The Solvay process

Three different products are formed from this process, light sodium carbonate (light ash), granular sodium carbonate (heavy ash) and refined sodium hydrogencarbonate.

The main by-product is calcium chloride. A little of this can be sold for use in refrigeration, curing concrete and as a suspension in oil drilling. The bulk of it is disposed of in the nearby river Weaver.

Light sodium carbonate (light ash)

This is made by taking the filtered sodium hydrogencarbonate and heating it. This drives off the water and the carbon dioxide, which can be recycled. The product is a very fine white powder.

Granular sodium carbonate (heavy ash)

Light sodium carbonate is made into a slurry with water, to form sodium carbonate monohydrate. This is then heated to produce the anhydrous form as much larger crystals. These crystals have a particle size similar to that of sand so that the two mix easily, which is important for glass making – the major use for heavy ash.

Refined sodium hydrogencarbonate

Crude sodium hydrogencarbonate is filtered and decarbonated and dehydrated by heating to give sodium carbonate. This is dissolved in water and the resulting solution is filtered to remove impurities. Highly pure crystals of sodium hydrogencarbonate are formed by reacting the filtered solution with carbon dioxide. These crystals are then centrifuged and dried in a carbon dioxide atmosphere. The product is one of the purest industrial chemicals and can be added to foods and pharmaceutical products.

This page has been intentionally left blank.

RS•C

Part 5
The thermodynamics and equilibria involved in the Solvay process for producing sodium carbonate

Material for post-16 students

The thermodynamics and equilibria involved in the Solvay process for producing sodium carbonate

Thermodynamic aspects of the process

This is an old process dating from the late 19th century. It is also known as the ammonia-soda process. It uses two raw materials: sodium chloride and calcium carbonate.

The overall reaction:

$$2NaCl(aq) + CaCO_3(s) \rightleftharpoons Na_2CO_3(aq) + CaCl_2(aq)$$

is endothermic ($\Delta H = +20$ kJ mol^{-1}, $\Delta G = +60$ kJ mol^{-1})

and the equilibrium lies well to the left. So the production of sodium carbonate must be undertaken by an indirect route. The actual series of reactions used is:

1) $CaCO_3(s) \rightarrow CaO(s) + CO_2(g)$ $\Delta H = +178$ kJ mol^{-1}

2) $2NaCl(aq) + 2NH_3(aq) + 2H_2O(l) + 2CO_2(g) \rightleftharpoons 2NH_4Cl(aq) + 2NaHCO_3(s)$
                                                                  $\Delta H = -158$ kJ mol^{-1}

3) $2NaHCO_3(s) \rightarrow Na_2CO_3(s) + CO_2(g) + H_2O(l)$ $\Delta H = +85$ kJ mol^{-1}

4) $CaO(s) + H_2O(l) \rightarrow Ca(OH)_2(s)$ $\Delta H = -65$ kJ mol^{-1}

5) $Ca(OH)_2(s) + 2NH_4Cl(aq) \rightarrow CaCl_2(aq) + 2NH_3(aq) + 2H_2O(l)$

                                                               $\Delta H = -20$ kJ mol^{-1}

Questions

1.   *(a)*   Show that the overall effect of the five reactions above is the same as:

                      $2NaCl(aq) + CaCO_3(s) \rightleftharpoons Na_2CO_3(aq) + CaCl_2(aq)$

You can do this by adding up the five equations above – *ie* adding all the species on the left of the arrows, adding all the species on the right of the arrows and then cancelling all the species that occur on both the left and the right.

      (b)   Hess's Law states that the enthalpy change for any reaction is independent of the route by which that reaction occurs. So adding the enthalpy changes of the five reactions above should give the enthalpy change of

                      $2NaCl(aq) + CaCO_3(s) \rightleftharpoons Na_2CO_3(aq) + CaCl_2(aq)$

Add up the enthalpy changes of reactions 1 – 5 above to calculate a value for ΔH for the overall reaction. Comment on the value you obtain.

The overall reaction requires an input of heat energy. This is provided by burning coke.

   $C(s)$ $+$ $O_2(g)$ $\rightarrow$ $CO_2(g)$

2.   What is coke and where does it come from? What other products are formed in the coke-making process?

3.   Use the table of data at the end of this sheet to find $\Delta\bar{H}$, the standard enthalpy change for the burning of coke.

The heat energy produced is used to decompose calcium carbonate (limestone, $CaCO_3$) according to the following equation:

$$CaCO_3(s) \quad \rightarrow \quad CaO(s) \quad + \quad CO_2(g)$$

Questions

4.   Use the table of data at the end of this sheet to calculate $\Delta\bar{H}$, the standard enthalpy change for the above reaction.

5.   Using the two $\Delta\bar{H}$ values you have obtained above, deduce (to the nearest whole number) how many moles of calcium carbonate can be decomposed by the energy produced on burning 1 mole of carbon (coke).

6.   Burning coke is the preferred energy source for the Solvay process rather than other fossil fuels or even electrical heating. Look at the set of reactions which make up the Solvay process and suggest why coke is used.

Coke and limestone are burned together in a kiln and the resulting carbon dioxide is used in the next stage of the process.

Sodium chloride solution (brine) is extracted from underground deposits by 'solution mining'. This involves pumping water down into the salt strata. The salt dissolves, and the saturated brine is pumped to the surface. Here it is purified and mixed with ammonia before entering a tower – called the Solvay tower – where the the reaction with carbon dioxide takes place, *Figure 1*. Here it reacts with the carbon dioxide according to the following equation:

$$NaCl(aq) + NH_3(aq) + H_2O(l) + CO_2(g) \rightleftharpoons NH_4Cl(aq) + NaHCO_3(s)$$

$$\Delta H = -79 \text{ kJ mol}^{-1}$$

Sodium hydrogencarbonate is less soluble than ammonium chloride at low temperatures. It precipitates out and is separated by vacuum filtration. Since this is an exothermic reaction careful temperature control is necessary to keep the temperature low enough for the sodium hydrogencarbonate to precipitate, *Figure 1*.

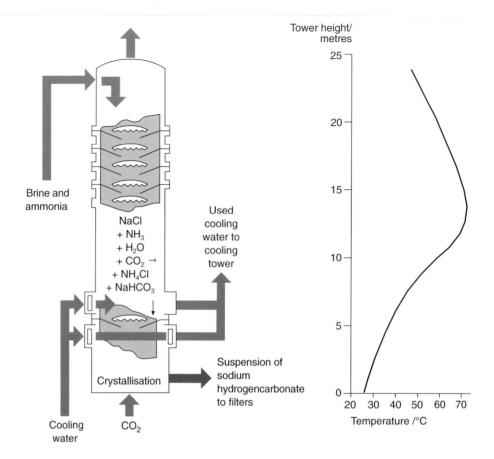

Fig 1 The temperature profile of a Solvay tower

Questions

> 7. (a)   Describe the advantages in obtaining salt from underground
>   deposits, up to 500 m deep, by solution mining rather than by
>   conventional mining.
>
> 8.   Explain the shape of the graph of temperature against tower height in
>   *Figure 1.*
>
> 9.   The crystal size of the final product is important. When crystallising a
>   product, what factors determine the size of the crystals produced?

The solid sodium hydrogencarbonate is transferred to driers, called secheurs, where
thermal decomposition takes place to form sodium carbonate.

Questions

> 10.   Write a balanced equation and calculate the standard enthalpy change for
>   the decomposition of sodium hydrogencarbonate using data from the table
>   at the end of this sheet.

Question

11. Look at *Figure 2* and describe in words or in chemical equations the processes that are occurring in the areas labelled A, B and C? What substances are represented by the arrows labelled a, b, c and d?

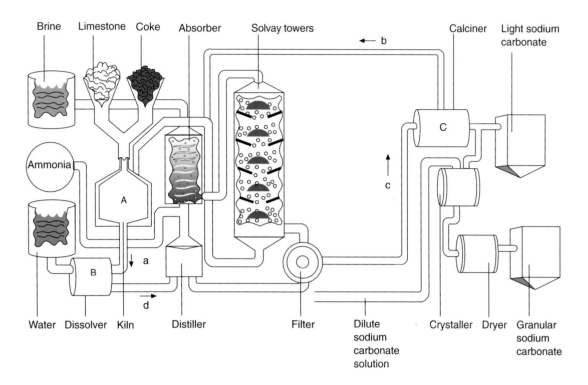

Fig 2 The Solvay process

Acid and base aspects of the process

The Brønsted Lowry definitions of acids and bases are that an acid is a proton (H^+ ion) donor and that a base is a proton (H^+ ion) acceptor.

Questions

12. Using the definition of acids and bases given above, identify the acids and the bases in the following reactions from the Solvay process.

   (a) $NaCl(aq) + NH_3(aq) + H_2O(l) + CO_2(g) \rightleftharpoons NH_4Cl(aq) + NaHCO_3(s)$

   (b) $Ca(OH)_2(s) + 2NH_4Cl(aq) \rightarrow CaCl_2(aq) + 2NH_3(aq) + 2H_2O(l)$

13. Sodium carbonate is the salt of a strong base and a weak acid.

   (a) Will a solution of sodium carbonate be acidic, neutral or alkaline? Explain your answer.

   (b) (This part is much harder)

   What will be the pH of a 1 mol dm^{-3} solution of sodium carbonate?

   This involves the two equilibria

   1. $CO_3^{2-}(aq) + H_2O(l) \rightleftharpoons HCO_3^-(aq) + OH^-(aq)$

   $K_b = 2.08 \times 10^{-4}$ mol dm^{-3}

   and

   2. $HCO_3^-(aq) + H_2O(l) \rightleftharpoons H_2CO_3(aq) + OH^-(aq)$

   $K_b = 2.2 \times 10^{-8}$ mol dm^{-3}

   You will first have to use the expression

   $$2.08 \times 10^{-4} = \frac{[HCO_3^-(aq)]_{eq}\,[OH^-(aq)]_{eq}}{[CO_3^{2-}(aq)]_{eq}}$$

   to calculate $[OH^-(aq)]$ produced by equilibrium 1,

   then you will need to use the expression

   $$2.22 \times 10^{-8} = \frac{[H_2CO_3(aq)]_{eq}\,[OH^-(aq)]_{eq}}{[HCO_3^-(aq)]_{eq}}$$

   and the $[HCO_3^-(aq)]$ you have worked out above to calculate

   $[OH^-(aq)]$ produced by equilibrium 2.

   The total $[OH^-(aq)]$ will be the sum of these two values.

   You will then need to use the expression

   $[H^+(aq)]_{eq} \times [OH^-(aq)]_{eq} = 1.0 \times 10^{-14}$ mol^2 dm^{-6}

   to calculate the corresponding value of $[H^+(aq)]$ and hence the pH of the solution.

Uses of sodium carbonate

The major use for sodium carbonate is in manufacturing glass. Here, sodium carbonate is heated with sand and other materials. Often the sodium carbonate supplied can contain an impurity of sodium chloride and this produces hydrogen chloride gas during the glass making process which pollutes the environment. So it is important that the sodium carbonate does not contain a high level of sodium chloride.

Question

14. Using the information from the titration curve in *Figure 3* and the information about some indicators in Table 1, plan in detail a method for determining the purity of a sodium carbonate sample obtained from the manufacturers.

If possible, check your method with your teacher and carry out an analysis of a sample of sodium carbonate.

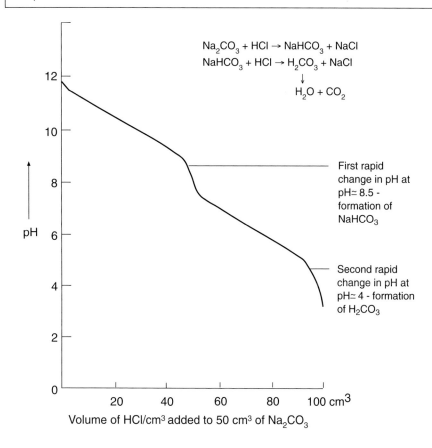

$$Na_2CO_3 + HCl \rightarrow NaHCO_3 + NaCl$$
$$NaHCO_3 + HCl \rightarrow H_2CO_3 + NaCl$$
$$\downarrow$$
$$H_2O + CO_2$$

First rapid change in pH at pH $\simeq$ 8.5 - formation of $NaHCO_3$

Second rapid change in pH at pH $\simeq$ 4 - formation of H_2CO_3

Volume of HCl/cm³ added to 50 cm³ of Na₂CO₃

Fig 3 Titration curve for sodium carbonate (0.100 mol dm⁻³) with hydrochloric acid (0.100 mol dm⁻³)

Indicator pH	1	2	3	4	5	6	7	8	9	10	11
Thymol blue	Red	Change			Yellow				Change	Blue	
Methyl orange		Red		Change	Yellow						
Methyl red				Red	Change		Yellow				
Litmus						Red		Change	Blue		
Bromothymol blue					Yellow		Change	Blue			
Phenolphthalein							Colourless			Change	Red
Universal indicator		Red		Orange	Yellow		Green	Blue		Violet	

Table 1 The colour changes of some indicators

Data

$\Delta H_f^{\ominus}$ /kJ mol^{-1}

$CO_2(g)$	-393
$CaO(s)$	-635
$CaCO_3(s)$	-1207
$H_2O(l)$	-286
$NaHCO_3(s)$	-951
$Na_2CO_3(s)$	- 1131

RS•C